Date: 2/22/12

**J 636.9766 RAN
Randolph, Joanne.
My friend the ferret /**

Published in 2011 by Windmill Books, LLC
303 Park Avenue South, Suite #1280, New York, NY 10010-3657

Copyright © 2011 by Windmill Books, LLC

All rights reserved. No part of this book may be reproduced in any form without permission in writing from the publisher, except by a reviewer.

First Edition

Editor: Jennifer Way
Book Design: Erica Clendening
Layout Design: Julio Gil

Photo Credits: Cover, p. 12 © www.iStockphoto.com/Fülöp Gergely; p. 4 Diane Macdonald/Getty Images; pp. 5, 6, 7 (bottom), 8, 10, 11, 15 (top, bottom), 16, 17, 19 Shutterstock.com; p. 7 (top) Tim Graham/Getty Images; pp. 9, 21 © Juniors Bildarchiv/age fotostock; p. 13 Getty Images; p. 14 © Mendil/age fotostock; p. 18 © www.iStockphoto.com/Xseon; p. 20 © www.iStockphoto.com/Leslie Banks.

Library of Congress Cataloging-in-Publication Data

Randolph, Joanne.
 My friend the ferret / by Joanne Randolph.
   p. cm. — (Curious pet pals)
Includes index.
ISBN 978-1-60754-974-1 (library binding) — ISBN 978-1-60754-980-2 (pbk.) — ISBN 978-1-60754-981-9 (6-pack)
1. Ferrets as pets—Juvenile literature. I. Title.
SF459.F47R36 2010
636.976'628—dc22
                                        2010004693

Manufactured in the United States of America

For more great fiction and nonfiction, go to www.windmillbooks.com

CPSIA Compliance Information: Batch #BW2011WM: For Further Information contact Windmill Books, New York, New York at 1-866-478-0556.

Contents

The Curious Ferret 4
Ferret Care 14
A Great Pet 20
Guess What? 22
Glossary 23
Read More 24
Index 24
Web Sites 24

The Curious Ferret

Have you ever wanted a furry, fun, **curious** pet? You might want a pet ferret, then!

You may think of ferrets as wild animals. There are ferret relatives that live in the wild. However, people have **domesticated** ferrets in the same way that cats and dogs

Ferrets can make great family pets. It is important to learn as much as you can about any pet before you get one.

have been domesticated. In fact, many scientists believe that ferrets were first domesticated more than 4,000 years ago. That's about 500 years before cats were domesticated!

Ferrets can make great pets. They are **social** animals. This means they like to be with other ferrets or other animals.

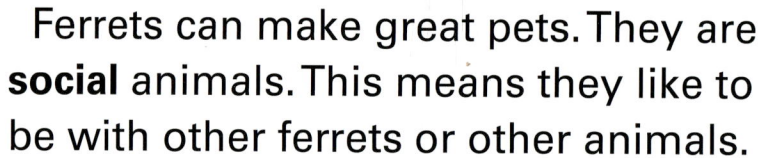

Ferrets like other ferrets, but they also get along with other animals, like cats and dogs.

If you plan to get only one ferret, then this "other animal" is you!

Ferrets are curious and playful. They can easily become bored. They may try to escape or chew on things if you do

Ferrets like the same kinds of toys cats like. They also like tubes and pipes they can crawl through.

not give them enough to do. Your ferret will enjoy having lots of toys and **tubes** to **explore**.

Ferrets play all kinds of games with each other. They like to dig, run, play tug-of-war, and explore new places.

TOOTH

Ferrets have sharp teeth. It is important to train your ferret not to nip during play so that it will not hurt you.

They love to crawl into small spaces. Sometimes, however, a ferret just wants to snuggle and sleep.

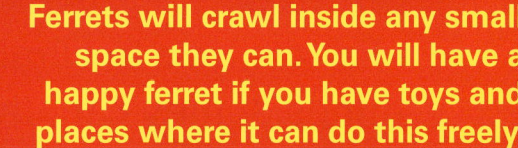
Ferrets will crawl inside any small space they can. You will have a happy ferret if you have toys and places where it can do this freely.

Baby ferrets may nip a lot as part of their play. These tiny bites can hurt. It is important to train ferrets to stop this behavior so they can play without hurting people or other animals.

Ferrets love to climb into and under things. If they can, they will crawl under your stove or refrigerator. They might

Ferrets like to hide under the couch.

try to climb into your dishwasher or your dryer. They can fit into very small spaces.

This baby ferret has curled up inside a vase.

Because ferrets like to explore such small spaces, it is important to watch your ferret closely. Ferrets can get hurt if they are hiding inside something and those things are turned on or moved.

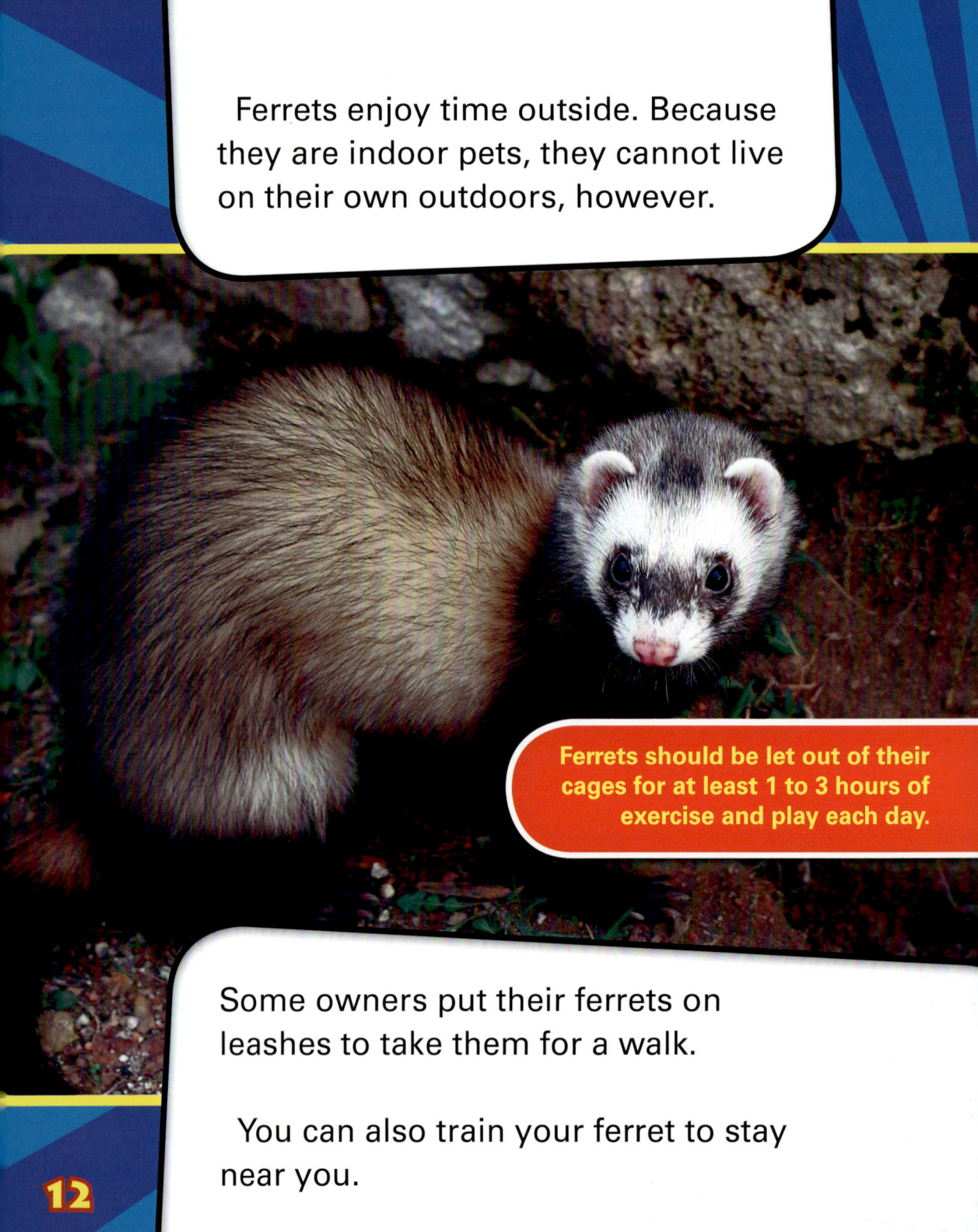

Ferrets enjoy time outside. Because they are indoor pets, they cannot live on their own outdoors, however.

Ferrets should be let out of their cages for at least 1 to 3 hours of exercise and play each day.

Some owners put their ferrets on leashes to take them for a walk.

You can also train your ferret to stay near you.

Only let your ferret off its leash if it is well-trained or if your yard has a fence that ferrets can't get through, over, or under.

This takes work, but can be rewarding. Your ferret will get to enjoy its freedom outdoors. You will feel comfortable that it will not get lost.

Ferret Care

Ferrets are very busy animals. Even ferrets need to sleep, though. In fact, ferrets can sleep up to 20 hours a day! They need a safe place to be during these hours.

This is a vet, or animal doctor. He is weighing a ferret to make sure it is healthy.

Baby ferrets need to sleep more than grownup ferrets do. When they are awake, though, they need to be trained.

You can buy a cage for your ferret at a pet store. You can put food, water, and a **litter box** there. Ferrets also like dark, closed spaces. Putting a sleeping box in your ferret's cage will help it feel safe.

Some people hang hammocks in their ferret's cage. Ferrets can nap happily there.

Ferrets have beautiful, soft fur. Their coats can come in different colors. The fur can be a solid color, or have markings on it. Some of the most

Here you can see two ferrets with different fur colors. The one on the left is dark-eyed white, and the one on the right is sable.

common colors for ferrets are sable, cinnamon, albino, chocolate, silver mitt, and butterscotch.

> You will need to bathe your ferret sometimes. Many ferrets do not like baths, but if you start giving them one as babies, they will get used to it.

You will need to bathe your ferret sometimes. Ferrets are related to skunks. This means their fur can smell bad sometimes. Keeping their fur, bedding, and cage clean can cut down on this odor.

Ferrets need to eat often. They should always have food in their bowls. They should always have plenty of clean water, too.

Ferrets love raisins. However, any treats, such as chicken, beef, or peanut butter, should be given in very small amounts.

Ferrets need healthy food with the right **nutrients**. There are foods made just for ferrets. However, they can eat

Ferrets eat small amounts of food throughout the day. That is why it's important to make sure your ferret always has food and water.

other things, such as dry cat food. Dry dog food and wet dog and cat food are not good for ferrets, though.

A Great Pet

For many years, some people were afraid of ferrets. They did not think they made good pets. It used to be against the law to own ferrets in many states.

However, thousands of years of experience may say otherwise.

Today ferrets can be found in most pet stores. They can make a great new friend for your family!

Ferrets are playful animals. It's fun to watch them playing with one another.

The millions of people who own ferrets would likely say otherwise, too. In fact, today, ferrets are the third most–liked pets after dogs and cats!

GUESS WHAT?

Ferrets generally live between 6 and 10 years. Baby ferrets should not be adopted until after they are 8 weeks old.

There are around 8 to 10 million ferrets living as pets in the United States.

A ferret has a long head that is flat on the top. This is a good shape for helping them to nose their way into tight spaces.

You need to teach your ferret to use a litter box. You do not want your ferret to choose its own place. After all, it could choose your bed or your shoe!

If you bathe your ferret too often you will actually make it smell more. Its body will work to make more of the smelly oil that coats its fur.

GLOSSARY

CURIOUS (KYUR-ee-us) Interested in new things.

DOMESTICATED (duh-MES-tih-kayt-id) Raised to live with people.

EXPLORE (ek-SPLOR) To go over carefully or move over little-known land.

LITTER BOX (LIH-ter BOKS) A box filled with matter that takes in wetness and that small animals use as a bathroom.

NUTRIENTS (NOO-tree-unts) Food that a living thing needs to live and grow.

SOCIAL (SOH-shul) Living together in a group, or enjoys being in a group.

TUBES (TOOBZ) Something that is long and has a small opening.

Read More

Hamilton, Lynn. *My Pet Ferret*. New York: Weigl Publishers, 2010.

McKimmey, Vickie. *Ferrets*. New Jersey: TFH Publications, 2007.

Morton, E. Lynn. *Ferrets*. New York: Barron's Educational Series, 2000.

Index

B
behavior, 9

C
cage, 14–15, 17

D
domesticated, 4–5

F
fur, 16–17

Web Sites

For Web resources related to the subject of this book, go to: www.windmillbooks.com/weblinks and select this book's title.